Top 30 Seeds for Pocket Edition

Steve Ranger Jr.

Copyright © 2017
All rights reserved.
ISBN-13:978-1978335141:

Introduction

Minecraft is one of the most successful videogames in history. Many think this is due to the wide range of options offered by Mojang, the developers, and the modding community. Nevertheless we think that the beauty of Minecraft lies in the randomness of the environment, the feeling of utter solitude, the vastness of the landscape, and, consequently, the uniqueness of each adventure.

The Random Number Generator (RNG) has much to say about this. The sceneries that we enjoy in our gameplay is not preset. What we see and where we walk is decided mathematically by complex functions. The core of this method is the seed, a number chosen by fate at the moment you decide to create a new world. The number of available seeds is so big that is hard to express without biting your tongue. 18446744073709551616. Sadly, only a small fraction of them are eligible when you create a world from your launcher.

This fact is the reason behind our book. There are amazing seeds out there waiting for you, but you cannot play them unless you know the secret code to generate them.

Minecraft seeds are unique for each edition and version of the game. Differences among consoles are inexistent in terms of the seed algorithm. Both Minecraft 1.21 for the PlayStation systems, Minecraft TU30 for Xbox 360 and Minecraft CU18 have interchangeable seeds. These are different from the ones found in Minecraft 1.8.8 for PC as they use the programming language C++ instead of the infamous Java for the RNG. Minecraft PE 0.13.0 is a world in itself.

As far as the world generator does not change, the seeds keep being useful after an update. The seeds we share here work for the versions stated above, but may break in the future. Minecraft 1.9.1, due to early 2016, will be compatible with the seeds coming up next..

CONTENTS

CONTENTS V

1 CHAPTER 1

MINECRAFT 1.8.8 FOR PC

2 CHAPTER 19

MINECRAFT PORTABLE EDITION 0.13.0

3 CHAPTER 33

MINECRAFT CONSOLE EDITION TU30 FOR XBOX 360

4 CHAPTER 37

MINECRAFT CONSOLE EDITION 1.21 FOR PLAYSTATION 3

5 CHAPTER 40

MINECRAFT CONSOLE EDITION CU18 FOR XBOX ONE

6 CHAPTER 43

MINECRAFT CONSOLE EDITION 1.21 FOR PLAYSTATION 4

CONCLUSION **47**

ABOUT THE AUTHOR **48**

1 CHAPTER

MINECRAFT 1.8.8 FOR PC

The computer edition is the most advanced at any time. The world border is hardly reachable by normal means. The surface of the word is equivalent to three thousand and six hundred million square kilometers, considering that each block measures one meter. All the seeds below belong to the normal terrain generator. Amplified worlds, large biomes and presets work differently so one must keep an eye on the options before generating the world.

World's Most Amazing Seed: The Infinite Mineshaft

SEED:107038380838084

You may enter the word and think something feels wrong. You still do not know what is going on, you can only see some trees, a nice beach and plains. Everything you need to start a small base is within the island.

The irk can only grow. When you go underground something feels odd. You sail to the North till you arrive to a new continent. It is cold, the pines are scattered all around and you can also see some pumpkins. You decide to settle in x=500, z=-1000.

The overwhelming surprise hits you when spelunking. While you mine your way down you start to hear monster noises galore. You try to find the place dreaming for some precious minerals. But you find the biggest adventure yet. An infinite mineshaft that goes from one border of the world to the opposite one.

Explore the humongous maze system as you advance towards the Farlands, retrieve the ancient glamour of the mines, set multiple cave spider grinders in the area or burn all down to ashes. This glitched seed originally discovered by Azelef is now yours. Use it wisely.

A Seed Selected Among Billions: Quad Witch Hut Seed.

SEED:91017135969371

When their server begun to stall in 2013, ZipKrowd members, some of the most brilliant minds in Minecraft, decided to start anew. To avoid past mistakes they planned every detail in advance. The world renowned redstoners thought that the hyper efficient witch grinder they had designed wasn't yielding enough redstone and glowstone.

As a consequence they begun a witch hunt of their own. They first interpreted the Minecraft structure generation code finding certain circumstances under which four huts could spawn close enough to make a much bigger farming facility. Soon after, their coders created a program capable of analyzing seeds automatically. Then they went to share the application with their fanbase so their computational power grew exponentially. Among the several billions seeds studied, only one was chosen.

Unluckily these seeds don't work as of Minecraft 1.8.8. Nevertheless a follower of the server took on the task to find a similar one for the current version. Thanks to NTBBPG we can enjoy a magnificent world.

The spawn is located in the border of a mixed forest. From there, we can head towards x=150, z=1750 to find an interesting location.

It's an almost landlocked mushroom biome of great size. Quite a sight. Mooshrooms wander over the mycelium and amazing fungi inhabits the place. The biome is bordered by the equally rare mesa.

The place is peaceful, and a great location to build a home. The underground is filled with resources. An astonishing curved ravine of epic dimensions is located at x=-70, z=1700, it carves into de ground reaching diamond depths. In fact if you inspect the area carefully you will find some there. Remember that under the Mushroom biome everything is calm, the only threat come from dungeons, lava and falls.

Top 30 Seeds. Ultimate Pocket Edition

At x=-1100, z=-600 you can find the quad witch huts. You will need a lot of skill, determination and knowledge to turn this swampy area into a high tech grinding machine. We can only cheer you if you take on this challenge.

This world also features other interesting biomes not too far away from the spawn. There is a hidden mesa Bryce, but what is more interesting is the presence of a jungle at x=800, z=-800. From there you can maybe spot a jungle temple. If your adventurer spirit guides you there, you will find some shiny loot.

If what you find isn't good enough for you, then the treasure located under the water at x=300, z=300 might be more interesting. Just be careful, there are dangerous monsters around.

All Biomes Seed: Minecraft in a Nutshell

SEED:-9223372036547736247

Steve Ranger Jr.

Few weeks after the launch of Minecraft 1.7.1 The Update that Changed the World our fellow minecrafters were already fed up of not being able to find some rare biomes. In order to satisfy the needs of every players, a group effort started to find a seed that had all biomes within a two thousand block radios from x=0, z=0.

This was accomplished little after. But why should you conform with having those places as far when you could compact them even more. This is what encouraged user foreversnoopy to search for this seed. It only took over 1500 hours of computational time to find it...

The spawn features the rare Forest Hills which have great resemblance with how Minecraft looked back in the beta days. Not far away a field full of sunflowers announces the change of biome. The Sunflower Plains area around x=-300, z=125 is a bucolic place to settle down and build up some sheep herds. The amazing Ice Spikes biome is located at x=400, z=500. There is also two Savanna Plateau Mutated biomes within range, the one at x=1000, z=-150 surpass the clouds reaching an outstanding height of 149 blocks at the summit. A little further from the limit, a nice Savanna Plateau descends towards a ravine, shaping a grin in the terrain. Certainly spooky. At x=-350, z=-250 you can enjoy the slim birch trees growing in another mutated biome.

The whole seed oozes with exploration.

Top 30 Seeds. Ultimate Pocket Edition

Diamantiferous Seed: Full Diamond Gear with No Effort

SEED:1002221280

Mining for diamonds can be a dangerous, tedious and a boring task. Happily we got you covered with this amazing seed originally showcased by youtuber QuiqueJaraGamer. The spawn is located in a cluster of warm biomes. There are deserts everywhere, and so are desert temples. At x=50, z=50 your first six diamonds are waiting in the secret chamber of the pyramid. Going to x=275, z=-225 will yield five additional diamonds. Two more await at a jungle temple located at x=525, z=-200. The stronghold at x=-775, z=425 brings the total diamond count to eighteen. Another desert temple at x=-250, z=225 adds three more and if you don't mind mining a little. Making a 20x20 room at y=6 below the center of the temple's secret chamber will provide ten diamantiferous ores. At this point you can make a full armor, sword, shovel and pickaxe. But if you still miss your axe and your hoe, visit the Nether Fortress at x=-75, z=-100 that will suffice.

Another Weird Seed: Moldy Spawn and Incomplete Portal

SEED:-77449594

As soon as you spawn you see pines and red blocks everywhere. It's not Christmas yet. The reason for such a colorful spawn is the existence of a Mushroom Island connected to the mainland in the neighboring area.

Nor far from there is quite interesting ravine. There are diamonds, gold, lapislazuli... and the ceiling is made of prismarine. As you have probably guessed there is a Water Monument above. You can loot eight gold blocks, find a room full of sponges and if you dare, make an Elder Guardian your pet. This place is at x=375, z=650.

Top 30 Seeds. Ultimate Pocket Edition

When you are ready to go, craft a boat and explore the surrounding archipelago. There is a disjoint End portal room floating in the sea. The funny thing is searching for it and enjoying the little weird things Minecraft throw at us. If you insist in having the coordinates here they are: x=-200, z=750. You will need a lot of Ender Eyes to reach the end as there are none placed.

Thanks to our friend DonkeyPuncher76 for this amazing find.

Hot and Cold Seed: Biomes Conflicting

SEED:1826678330

The spawn is a beautiful Flower Forest surrounded by a river whose wellspring is in the nearby mountains.

The really striking landscape of this seed can be enjoyed at x=775, z=-575 where two opposite biomes clash and form amazing sceneries, impossible overhangs and a valley of antithesis. Skilled builders will find in the complicated relief an interesting challenge.

There are two other places worth mentioning. The first is the Mega Spruce Taiga at x=1150, z=-775 where you can tame a wolf to get a companion. The last is the meeting of a mesa Bryce with the very same jungle (which spans for well over a thousand blocks in each direction). This place can be found at x=1200, z=700.

This seed found by AWEDYSSY sure has a lot of potential for both adventurers and builders, we are eager to know what our reader will make with it.

Amplified Seed? Not Really

SEED:82735704725346o1379

Redditor xBcrafted found a seed with strong reminiscence of the old Minecraft days, a seed that has the best of both amplified and normal worlds. Your adventure starts here in a calm savanna where you can find cows and sheep. If you go ahead you will feel like an ant as a massive massif stands imposing in front of you.

There are arches, overhangs and strange shapes that will remind you of the real speleothems in our caves. If you manage to reach the top, you will also see internal valleys and further mountains above the clouds, as well as a village down below!

If you haven't already seen it, there is a few houses East to the spawn. You can get carrots, potatoes and wheat in their crops, as well as enough obsidian for an enchanting table. If only you could find some...

The village has a river. If you follow the riverside to the North till you reach the desert, you will find a temple there and some reeds. The diamonds that you're missing are there, try not to activate the TNTs or everything will be blown up!

Now you can enchant yourself some iron gear. There is plenty of it anywhere underground.

Survivalist Challenge Seed: A Village in the Sea

SEED:-6061212247400511253

Survival islands are one of the most demanded seeds. We didn't want to share a barren island without any appeal to it, so here is an island with a twist.

Seconds after you log in the world you will be hearing the annoying noises of your new neighbors. In front of you there is a blacksmith, but don't get too excited, there's not much to loot. Instead, the most valuable resources around are the crops.

This is an awesome seed to play in hardcore mode. How much will you last? How far can you develop your new empire? Will you survive without the need of leaving the island boundaries. There is food, water, animals, trees, water again...

When you are ready for some mining, head to the lava lake, dig some stairs downwards till you reach a large ravine. You will be welcomed by ferocious mobs and shiny rocks.

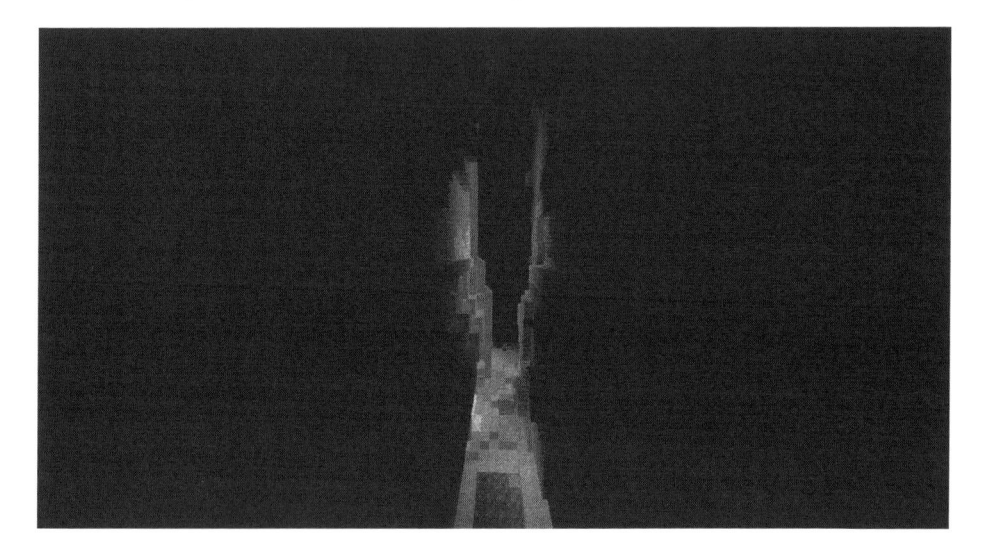

This seed discovered by CrashFilmsHD has nothing aside a lonely island in a vast ocean. A true challenge for a skillful player.

Our Awesome Reader's Seed: Thank You!

SEED:-645830912

As you might have noticed in the seed blank space you can enter any character. Those are later transformed into a number which is the numeric value used for all the pertinent calculations. In our case we input "Our Awesome Reader's" and retrieved the numeric seed using the command /seed (one of the very few available in Survival Mode),

You spawn in a calm mixed forest. Below your feet there is a complex system of caves, including several ravines next to each other. The area is not deep enough to be actually worth your time.

At x=-400, z=-750 you will find one of the many villages. This one has an armory where you can get some gear, Dark Oak Forests surrounding the area and a profound hole in the ground that leads to a cave that spirals down to the lava lakes. There is gold everywhere, a zombie dungeon at x=-480, z=-780, and we counted several dozens of diamonds, although they are well hidden in the stone.

The end portal is near to x=950, z=-100. It's part of a big stronghold located under an island. The labyrinth is oriented

northwards and contains at least seven chests. As soon as you reach the northernmost part you may hear the noises made by some venomous crawling critters. If you reach the end of the corridors and start some stairs at x=900, z=200 you will enter a gigantic mineshaft.

There are shiny rocks and monster spawners to cover your needs, including another zombie dungeon. The interlacing aisles go on and on and on. It can take several days to extract all the precious metals and crystals found in the place.

If your alchemist blood is urging you to brew some potions, The Nether is waiting for you. There is a compact Nether Fortress at x=-300, z=-100. There is mystical netherwart in one chest, a blaze spawner and a triple cross section that allows you to make a large wither skeleton grinder of it. Your seed has everything a Minecraft player may ever need.

Top 30 Seeds. Ultimate Pocket Edition

Triple Top Notch Seed: A Desert Full of Structures

SEED:-872384114397604234

You spawn in the savanna but as soon as you start moving you notice the desert. Before you go there you might want to grab some food at a village. There is one not far from the spawn, at x=-300,z=-300. Once you are ready head towards the West. There is another village at x=-900, z=-450 where you can replenish your resources. Then keep going to x=-1275, z=-700. If you need more food you can go to x=-1450, z=-400, if not go directly to x=-1800, z=-500. There is another interesting spot around x=-1950, z=300. Look well because it is pretty hidden.

Steve Ranger Jr.

Then go to x=-1925, z=1325; x=1700, z=825; x=-1300, z=600; x=-950, z=225; x=-400, z=250; x=-150, z=1050; x=200, z=1300; x=900, z=1125; x=340, z=545 and x=225, z=0. Your inventory should be similar to the picture below. We do the mathematics for you. You're lacking one apple to make three Notch Apples. Now go to punch some acacias at the spawn!

2 CHAPTER

MINECRAFT PORTABLE EDITION 0.13.0

Minecraft PE depends on a modified C++ language and consequently its seeds are not compatible with any other edition. Worlds are virtually infinite and the structure spawning algorithm works oddly at the moment. Be careful with the OS that your phone is using, depending on it the spawn can change (but the terrain will not). The spawns of the following seeds work without fail in Android systems.

Great Village Seed: Why Not Four?

SEED:can u follow me?

As soon as your view renders you will see a village that spans through the land and reaches the skyline. This seed was first discovered by Resul97. While the apparition of twin towns is not that scarce, what is exceptional about this one is that there are four villages conjoined.

You will have enough armor, tools, wood, food and obsidian to cover your basic needs. Furthermore this place has a high chance of naturally spawning iron golems due to the high NPC population and number of houses.

You can start your adventure here and move later, or settle definitely and grow a commercial empire with your new found friends.

Troll Seed: Laugh at Your Friends

SEED:1000921212

This one is not for you. This is to have fun with your friends, to shut up that guy that keeps telling he is the best Minecraft player that has ever existed or to play some antics... Eystreem Gaming shared a video of this seed with the community. His seed is special because it has a high chance of spawning the player in lava.

Challenge your friends not to die in this seed in hard difficulty for five minutes. They will jump on the challenge and cry in sorrow as soon as their feet touch the fiery magma beneath. You may even get some money out of them, bid for their lost and get rich quick. Jokes apart, this seed is also awesome to get a cocky player on their feet. It's impossible to be prepared for all the random events that Minecraft has ready for us and so we must be humble about our capabilities.

Quick Start Seed: Structures to the Rescue

SEED:1695763201

In Minecraft you might spend endless hours grinding for

resources. Growing crops, mining minerals and managing cattle takes a lot of time at the beginning. If you prefer to leave these activities for later, you can use this seed for a quick start.

From the spawn you will be able to see a village of decent size. There is a blacksmith which will allow you to get an iron helmet, some bread and a couple of ovens. There are enough vegetables around to feed an elephant, be sure to get only those that are ready to pick.

If you go to the border of the town you will get a satisfactory sight. Another village in the desert. There is some more carrots and potatoes there. Nothing too special, but if you keep looking at the distance, you will eventually find a twin temple. In their chambers you will get enough diamond for a pickaxe, gold, emeralds to trade with the villagers and iron for your armor.

If instead of heading directly to the first village from the spawn we go in the opposite direction following the plains, we will eventually arrive to the third and biggest village. This one also has a blacksmith with very useful loot.

From this point ahead, you surely can build some infrastructure to speed things up. While you do it, thank DAN LIVE YT for the awesome world you are playing.

Strangely Epic Seed: Moldy Villagers

SEED: Wizzz000

This seed was found by BlazingTornado. He soon realized that this seed was so epic that someone had to showcase it on internet, and here we are.

You spawn in a village which invades the moldy soil of a Mushroom Biome. This already seems epic, but wait, because there is much more. Besides the abundant food sources (crops and mooshrooms) this seed features a mythological animal.

An animal so rare that most Minecraft players think that it does not naturally exist. When they are found in servers people immediately think they are subject to a hoax. We are talking about the pink sheep. With a microscopic 0.164% chance to spawn, this is one of the most uncommon things you can find in the game, together with The Killer Bunny of Caerbanogg, the seven block tall cacti and the full diamond armored mobs (yes, they do exist).

Top 30 Seeds. Ultimate Pocket Edition

You will find these sheep in the plains in front of the village. You will not have trouble to locate them if you wander around looking for materials.

If you explore the Mushroom Biome you will notice an archipelago belonging to the biome. If you look at it from the coast and keep advancing past the islands, you will arrive to another fairly infrequent biome, a beautiful mesa.

Tenuous Crystals Seed: A Frosty Venice

SEED:kop

Youtuber JereVids submitted to his network a video with five awesome seeds, one of them is certainly outstanding, so we are including it in our compilation,

You start in a prairie with a village. Nothing special there besides the fact that the inhabitants have built bridges over the river to cross it. You will not see this in any other seed.

Furthermore, as soon as you take some time to look around, you will find an Ice Spikes Biome surrounding all the area. You won't ever need more compacted ice, you'd better grabbing your silk touch tools and getting some of it for your own constructions, systems and machinery.

This seed has a special feeling to it and we are pretty sure that you're going to enjoy your new white paradise.

Island Cluster Seed: Claim the Archipelago

SEED:thebestman

A cool seed to take on the traditional challenge. The spawn island isn't large enough, so you can find more few blocks away. But be careful, if you seek for new lands too far away, you will be engulfed by the vastness of the ocean.

Manage your resources wisely to enjoy the challenge. Start getting some wood and replanting the trees to keep the environment healthy. Collect some seeds and start a crop in the coast. Grow wheat for your nourishment needs and start thinking about connecting the islands.

Building a bridge to ease the transportation will be hard, but as soon as you finish it you can start farming monsters. Bones will increase your chances against hunger, while silk will allow you to better defend yourself thanks to a bow and fish some treasures with a baited rod.

If you are feeling confident in this seed provided by Captain Miah, you can even try to set a passive mob grinder. Then you could say you've actually conquered the lands and gained the honor to claims yourself Emperor of the Scattered Archipelago.

Jungle Seed: Introducing My Pet Cat Melon

SEED:-36599876

You will need a powerful device to run this one. Minecraft experts classify leaves as transparent blocks, these are very taxing on the CPU as Minecraft calls for redraws of the blocks behind them. Exploring this seed can bottleneck the resources of your phone and crash your game as there are leaves everywhere.

Getting a cat in Minecraft PE is not an easy job. Ocelots only spawn in jungle biomes, so you need to find one first. Luckily, you are reading the best Minecraft e-book available in the market, and we knew you would eventually need a cat.

You spawn in a jungle. there are no ocelots to be found but you can force them to appear. Melons are ubiquitous... wouldn't it be funny to name your tamed feline Melon? It surely is.

To make sure some ocelots spawn you will need to explore the underground fully. The free cats are considered hostile and share the cap with other monsters so, in order to make their spawn more probable you will need to light the caves and wait for the day. Burning up the whole place will help too! Now you got an excuse.

You can start by bordering the jungle to find a village inlaid in the rainforest, this will give you a boost in form of basic resources. As soon as you feel confident, start your quest for Melon.

This seed is brought to you by spiderman indonesia, a Minecraft player for South East Asia.

Dungeon Seed: Skeletons, Experience, Power!

SEED:-792906360

Minecraft 1.8.8 with its Spectator Mode allows a quick exploration of the underground. Sadly, this is not possible in Minecraft PE. This is the reason why even the most commons structures are very valuable in this edition of the game.

Here we share with you a unique seed. You spawn in the margin of some charming plains. The very block were you are standing is the key for your success. If you dig down a tunnel till you reach a cave you will be welcomed by an army of skeletons.

There is a skeleton dungeon there. If you are a resourceful minecrafter, get a pickaxe and take the buckets from the chest to start your grinder. This will allow you to get armor, bows, bones and experience while you do your homework!

Do not hesitate and choose this seed. Let's see how far can you reach.

Stronghold Seed: A Well to Another Dimension

SEED: 805967637

This is one seed we love to play again and again. Anytime anybody asks, we recommend this seed. It's dynamic, fun and very satisfactory. We learnt about it from a Brazilian player: Planet Android.

The scenery is blocked by some cobblestone wall standing in front of you. Yes, there is a village. Again. Do you know where to find the goodies? No? Why don't you try to jump into the well and dive to the bottom?

The mossy bricks will give away the presence of a stronghold below the town. Chests, libraries, silverfish... Problems and riches galore! Finding out the portal might need some persistence, as a prize, it already has two Ender Eyes placed.

Can you kill the EnderDragon in less than ninety minutes? It's a challenge!

Skyscrapers Seed: High Buildings and High Expectations

SEED:34568098

This seed was found by professional seed hunter ipodzgaming. You almost suffocate in a blacksmith wall when you first spawn. As soon as you ginger up from the initial surprise you will notice that the houses in this village are not your casual houses. This are skyscrapers. They go up through the wall of the mountain and reach a non negligible height.

If you dig down the village you will be eventually reach the stronghold. But first you may want to go exploring around to find some extra temples and villages, you can gather a significant amount of precious materials from them.

Have a nice time, this seed will surely provide enough fun for an adventurer like you!

3 CHAPTER

MINECRAFT CONSOLE EDITION TU30 FOR XBOX 360

Minecraft Console Edition, besides having different versions, is the same game. This means that seeds are interchangeable between consoles. Nevertheless, we have decided to keep seeds separated by console as updates will take more or less time depending on the company. Xbox 360 has a limited world size of 862x862 blocks.

Everything Seed: Cover All Your Needs

SEED:-1094874636

This seed was found by Teresa Huntington. She noticed that a seed with seven villages, two desert temples and six spawners isn't found that frequently.

Steve Ranger Jr.

Your spawn is located in the plains margin, next to a snowy coniferous forest. From this point you can already spot two villages, one in the desert and another in the plains.

Both of them lack a blacksmith, but the former has a surface spawner next to it. Go from the spawn area to the sandstone village and turn right when you reach it. Close to the river you will see a hole with collapsed sand. There is a zombie trap there and a couple of chests, one even includes a name tag!

If you had left the village behind and kept walking towards the end of the desert, the result would be the same! There is another exposed dungeon, this time you will have to face some skinny skeletons. There is some quality loot in there. When you're done, look around, there is another village hiding past the dunes.

If you go NW from this village you will find a temple, but before you arrive to that area, you will have to trespass two additional surface dungeons.

More interesting locations are scattered above the surface, including a blacksmith containing eight obsidian blocks. The stronghold can be accessed via x=-120, z=100, there are diamonds around too.

Extreme Survival Island Seed: 90% Water, 10% Fungi!

SEED:THOR

User GlorY King suggested to try this seed, and the result was worthy. Another challenge for those who find Minecraft too easy.

Top 30 Seeds. Ultimate Pocket Edition

You have a tree. You have some land. There is no more. Do you have what it takes to survive in this inhospitable islet? Oh! By the way, there is an absolutely amazing Mushroom Biome close to your new home... Enjoy!

Perfect Seed: Beautiful and Unique

SEED:bambi

This seed has a lot of cool biomes around the spawn. Whatever your goal is, you're going to achieve it if you select this world. There are snowy taigas, swamplands, plains, forests, deserts, oceans and jungles.

The prairies are cropped with vigorous horses. try to train and breed them to get strains with better attributes. There are several villages, caves and ravines around the spawn, but what makes this seed truly special is deep inside the jungle. At x=-375, z=-275 you will find an ancient temple covered in moss. Beware of the traps! And, surprise, there is another one at x=200, z=-275.

For those of you that enjoy turning up these structures into your home, this is a pretty cool seed. Thanks to ECKOSOLDIER for his amazing find.

4 CHAPTER

MINECRAFT CONSOLE EDITION 1.21 FOR PLAYSTATION 3

This version has the same characteristics as the previous. The only difference being when do Sony and Microsoft push their updates.

One Mile Long Rail Seed: Infinite Mineshafts in Console Edition

SEED:107038380838084

The unusual code behaviour discovered by Azelef has prompted a search for an analogue in console.

This seed has the same properties as the original. Chunks get repeated infinitely in one axis forming stripes underground. One of the chunks that gets replicated contains a mineshaft and so, a mining system covering the whole world gets created. You can access it from x=-200, z=220.

Diversified Seed: A Little of Everything

SEED:-809115099

Mountains, plains, forests, deserts... Take your choice and start to enjoy a seed notorious for its variety. At x=150, z=275 you will find a village with food and loot. Around x=250, z=325 a surface spawner allows you to grind spiders. A couple of chest have fancy stuff actually, right now you're able of riding an armored horse. You only need to find one.

At the distance you can spot a small Mushroom Biome, and a little closer a witch hut with easy to flood surroundings, good news for the technical fanatics.

Top 30 Seeds. Ultimate Pocket Edition

If you are planning some mining, there are interesting caves near the bedrock at x=250, z=375. In fact, in this precise coordinates, you can find a cluster of eight diamond ores.

A cool Nether Fortress stands imperturbable among the reddish rocks of the Minecraft hell. Its coordinates are x=50, z=75. You will be able to obtain gold and diamonds, as well as netherwart, blaze rods and luxury kits for your equine folks.

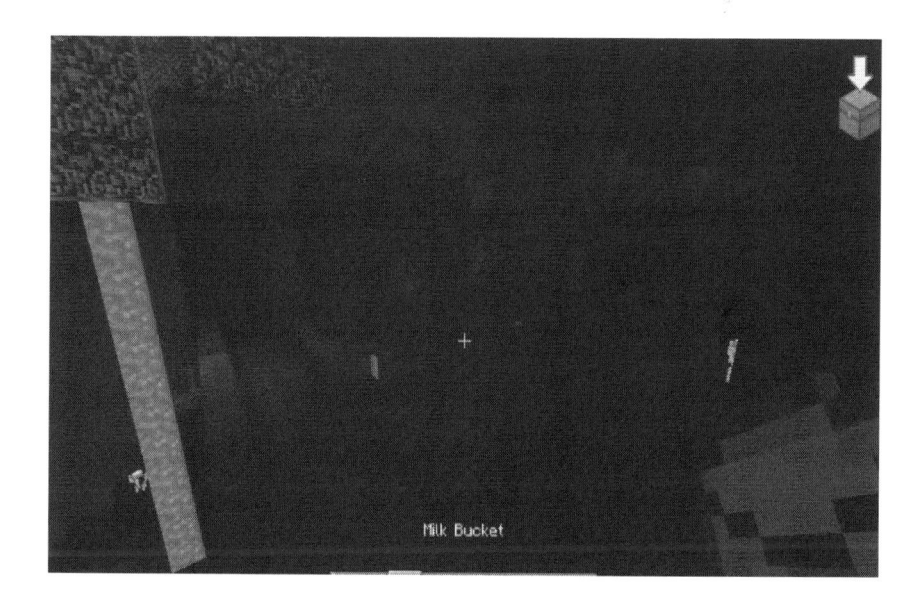

5 CHAPTER

MINECRAFT CONSOLE EDITION CU18 FOR XBOX ONE

The seed algorithm and RNG is the same used in the old consoles. The only difference being that the world spans for 5120x5120 blocks.

Massif Seed: A Life in the Mountains

SEED:37985039364542065

If you want to become a hermit you should start training now. This seed will spawn you in front of a landscape of touching beauty. Make a small shelter in the top of the mountain and descend to the desert village from time to time to get goods from the inhabitants in exchange for the emeralds from the undergrounds. This seed has also a jungle close to spawn, forests and taigas mark off the limits of the Extreme Hills and lakes give the whole lot a cozy aspect.

Top 30 Seeds. Ultimate Pocket Edition

Classic Seed: Why Did Notch Lie to Us?

SEED:-6042239240272384234

This is for seeking some nostalgia. In this e-book we have tried our best not to set different world options, but this one deserves it.

When strongholds where first announced to the public as an spoiler of the, then upcoming, Minecraft 1.8 Beta, it seemed like if they would generate at surface level allowing some bricks to stand out of the grass giving away the existence of the structure. That was not the case when the official beta version was finally launched.

However, if you set your world to Classic and enter this seed you will find two awesome things: a Mushroom Biome at x=-75, z=275 and a surface stronghold in the midst of a forest, located at x=60, z=205 where the leather covers of the library's books are cracking under the sun.

An awesome finding by Latha Math.

CHAPTER 6

MINECRAFT CONSOLE EDITION 1.21 FOR PLAYSTATION 4

Hilly Seed: Up and Down!

SEED:-7364770272654700515

If you want to start a new world that you could channel in different directions depending on the events, this seed is for you. There is a jungle around x=-50, z=200 that has several protruding mountains covered by overgrown vegetation. The summits extend towards the desert causing steep slopes there too. At x=-25, z=25 you have a fine example of what are we talking about. Following the same direction, you eventually arrive to a cluster of highlands plagued by emeralds, intriguing and whimsical shapes and life. It will be hard to build only in one of these places...

Easy Seed: A Seed for Beginners

SEED:162884248837880424

Every player must be a newcomer at some point. The sooner the better. This seed is for you. It has a terrain easy to walk through. The spawn is the corner among a swampland, some extreme hills

Top 30 Seeds. Ultimate Pocket Edition

and a jungle. This ensures that you get a selection of materials for your building and crafting needs.

Not far away a desert temple hides some loot. This place is at x=-325, z=-150. The center of the room hides a secret chamber. You can carefully staircase and get some iron and you first diamonds. If you feel unsafe you can craft them into a powerful sword. But the temple is not the only stuff to be found here. Next to it a strange hole in the sand is a clear clue to an experienced player. An exposed dungeon! You can assault it safely during the day.

Further away, you can spot a desert village. This is a source for easy food. Take what you need and replant the seeds. Another one of these can be visited if you reach x=-50, z=-700.

When you get the grip of the game, you can try harder adventures. Why not trying to explore a mineshaft located underground at x=-

100, z=-840? If you manage to survive nobody could call you a noob.

Your Seed: Random Adventure

SEED:

All the seeds gathered for the creation of this e-book are great, even exceptional or legendary. However, nothing beats the charm of your own adventure. One that no one else will be able to live. When you select a random seed, you are selecting nothing. It is Minecraft, the game, the one that chooses what challenges you will need to overtake, how the landscape will mould your enthusiasm, when will you be able to enter the other dimensions... All players must find that special seed that makes them happy. Just hit start a new world and don't look behind. Your Adventure starts now.

CONCLUSION

In this e-book we present thirty seeds that will cover all your needs. They are equally divided between the PC, Portable and Console editions. We want you to enjoy them as much as we enjoyed writing this e-book. If you do, please, do not forget to check our future publications.

ABOUT THE AUTHOR

Hi! I'm Steve Ranger Jr. (my real name is Steve Nebbit). I spent my childhood in San Diego, California, USA. I work as a developer of computer software at Microsoft.

Life is a game. Minecraft is the atmosphere of the game. There is a place for everyone here. Explore caves, steep to build structures, invite friends and build a village ... It's really cool and fun. "I think… I'm even sure that any person can change the world, and if you want a big change in the future then become this change today."

Printed in Great Britain
by Amazon